A Tree

Ruby Tuesday Books

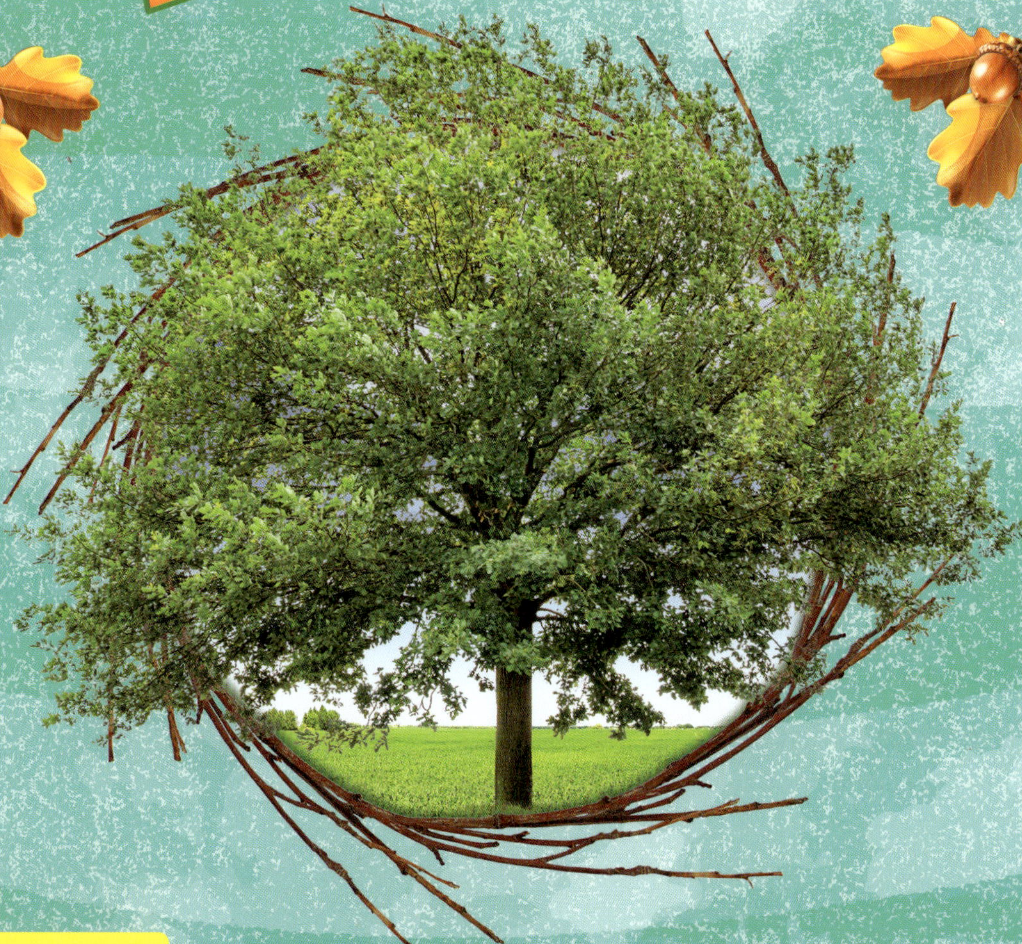

Ruth Owen

Published in 2025 by Ruby Tuesday Books Ltd.

Copyright © 2025 Ruby Tuesday Books Ltd.

All rights reserved. No part of this publication may be reproduced in whole or in part, stored in any retrieval system, or transmitted in any form or by any means, electronic, mechanical, photocopying, recording, or otherwise, without written permission from the publisher.

Editor: Mark J. Sachner
Design: Tammy West
Production: John Lingham

Photo Credits:
Shutterstock: Cover (Zerbor), 4 (George Trumpeter & Zerbor),
5, 6 (Mykhailo Pavlenko), 7, 8 (Lina7471), 9T (Dudarev Mikhail), 9B (Rodion Golova),
10 (Zerbor), 11 (Buslik), 12 (Yuri Antonenko), 13 (leonardo2011), 14 (New Africa), 15 (Valentina Razumova), 17 (David Savile), 18 (Valeriya Rychkova), 19 (Marilyn Barbone), 20 (Ohishiapply & Zerbor), 21 (Rudmer Zwerver), 22 (ABC photographs, Semmick Photo, & Willee Cole Photography), 23 (showcake, kaczor58, & Gabriele Rohde), 24 (Alexandr Shevchenko).

ISBN 978-1-78856-441-0

Printed in Poland by L&C Printing Group

www.rubytuesdaybooks.com

CONTENTS

Welcome to a New Day! 4

Glossary 22

Index 24

Welcome to a New Day!

It's morning, and the Sun is rising.

Cheep, Cheep, Cheep

Twitter, Twitter, Twitter

Tweet, Tweet, Tweet

Peep, Peep, Peep

A loud, happy noise comes from the leafy **branches** of an oak tree.

The birds that live in the oak tree are singing.

Their songs say
welcome to a new day!

Clouds float over the oak tree.

Rain falls from the clouds and soaks into the ground.

The tree's **roots** suck up rainwater from under the ground.

Roots

The oak tree drinks a bathtub of water each day.

The water flows from the tree's roots up its **trunk**.

Trunk

Bark

The tree's trunk is covered with **bark**.

The water flows along the tree's thick branches.

Branch

Twig

Leaf

It flows into thin twigs and into the leaves.

Now the tree's leaves have a special job to do.

They make sweet, sugary food for the tree!

The leaves make food with water, **gas** from the air

and sunlight.

The sweet food gives the tree energy.

New leaves

It uses the energy to grow new branches, twigs and leaves.

The tree also grows acorns.

Acorn

The acorns will turn brown and fall to the ground.

Last autumn, the tree's acorns fell to the ground.

Now the oak tree has a little neighbour.

It's a baby oak tree!

It is growing from one of the acorns.

An acorn is an oak tree **seed**.

The oak tree is like an animal town.

Baby squirrel

Squirrels live in a hole in the tree's trunk.

This hole is home to a woodpecker family!

All day, the tree makes food in its leaves.

It also helps make the air that we breathe.

When the Sun goes down, tiny night-time animals fly from the tree.

Bats!

Bat

Bats live in little cracks in the tree's bark.

The oak tree's busy day is over.

Glossary

bark
A tough, rough covering that protects a tree's trunk and branches.

branch
A long part of a tree that grows from the trunk. Twigs, leaves, flowers and fruits grow from branches.

gas
A substance in the air that we can't see. People and animals breathe gases in and out.

roots
Underground parts of a plant. Roots take in water and hold a plant in the soil.

Seed

New plant

seed
A tiny part of a plant that contains everything needed to grow a new plant.

trunk
The thick main stem of a tree.

Index

A
acorns 13, 14–15

B
bark 8, 21
bats 21
birds 5, 17
branches 4, 9, 12, 19

F
food 10–11, 12, 18

L
leaves 4, 9, 10–11, 12, 15, 18–19

R
roots 7, 8, 15

S
squirrels 8, 16

T
trunks 8, 16
twigs 9, 12